lfins   alvin lustig   sputnik   slinky

tv dinners   korean war   onk

senhower   saul bass   abstract expressionism

boomerang   color tv   ed sullivan   jazz

astics   jean arp   beatnik   buddy holly

civil rights   howdy doody   andy warhol

ster beall   jack kerouac   surrealism   kool-aid

supermarkets   barbie   atomic bomb   elvis

ol   pop art   game shows   the beatles

westerns   walter paepcke   paul rand

pering   william golden   biomorphic

In memory of my Grandmother, Agnes

Acknowledgments: Thanks to Grant Kessler for all the handsome photography. Ma
thanks to several friends for helping me edit text: Craig Keller, Carol Monagha
Audrey Tawa, and Jeff Burbank. Thanks once again to Bill LeBlond for his guidanc
and to those at Chronicle Books for their assistance on this project: J. Ada
Bluestein, Lesley Bruynesteyn, Patricia Evangelista, and Charles Robbins. The fo
lowing individuals have helped me with information, cleared copyrighted materia
with their companies, or supplied me with transparencies of their work: Craig Hur
at Jefferson Smurfit, Sharon Ptak-Miles at Kraft Foods, Sheree Clark and Joh
Sayles at Sayles Graphic Design, Robynne Raye at Modern Dog, Laurie Rosenwald a
Rosenwald Design, Bill Rawski at Zap Antiques and Props, John Mendenhall, an
Candice Chambers. I'd like to thank my brother, Ken, for giving me his boyhood rocke
transitor radio (still in the original box), and to my sister, Evy, for looking for vintag
packaging while in London.

ISBN: 0-8118-1784-9
Library of Congress Cataloging-in-Publication Data Available.

Editing: Charles Robbins
Text, book and cover design: Jerry Jankowski
Fonts: Univers and Bulmer

Printed in Hong Kong.

Distributed in Canada by Raincoast Books, 8680 Cambie Street, Vancouver, B.C. V6P 6M9

10 9 8 7 6 5 4 3 2 1

Chronicle Books
85 Second Street
San Francisco, CA 94105

Web Site: www.chronbooks.com

shelf space

modern package design

1945 – 1965

jerry jankowski

grant kessler/photography

CHRONICLE BOOKS

SAN FRANCISCO

shelf space

modern **package design**

1945–1965

introduction

Fifties and sixties packaging relied on a looser style of illustration and on photography to attract the eye and make the sale. In contrast, the Italian packaging for Audacia pen nibs and Presbitero pencils is quintessentially early thirties modern with its hard-edged design.

Aerosol sprays. Aluminum cans. Frozen dinners. Plastic bottles. They line the shelves in every American supermarket, in every Mexican super *mercado*, and in every Polish super *markecie*. Imagine a consumer world without them. But before World War II, none of these packaging innovations existed. Originally born out of the needs of the military, these and other new technologies were quickly adapted in the prosperity of peacetime that followed to create amazing new products to buy and alluring packages to sell them in.

After World War II, pent-up consumer demands and an increase in disposable income created an unimagined quantity and selection of things to purchase. Often proudly stated at the time was the belief that never before in the history of mankind were there so many consumer goods for so many to enjoy. Shelf space in supermarkets and other arenas of mass merchandising soon became high-priced real estate.

All this abundance was challenging for package designers. After all, how does one design a creative, aesthetically pleasing package when the client's boxed detergent is surrounded by a dozen other detergents all shouting at the potential buyer with their fluorescent colors and money-off coupons? In this new age of aggressive selling, could good design survive the aesthetics of plenty?

Packaging performs four essential tasks for any consumer product: contain, protect, identify, and sell. That product could be a canned consumable called Spam, a detergent "with super-wetting action" named Fab, or a fancy, perfumed powder christened Mais Oui. Packaging works for the low-brow and utilitarian product as it does for the high-brow and luxurious.

Types of packaging include boxes, cartons, cans, bottles, bags, and includes the labels applied to containers. And although every package is a container, not every container is a package because not all containers identify or help sell their contents.

Take the egg. An egg shell may be nature's perfect container, a well-designed form with an ingenious, tamper-resistant seal, but it still lacks qualifying information. Where did this egg come from? Is it really a chicken's egg? How fresh is it?

An egg is *not* packaging. Neither is an unmarked corrugated box, an unprinted paper bag, or an expertly wrapped birthday gift. However, a labeled, styrofoam container holding twelve hen's eggs is packaging. This particular package has the company's brand name, FarmFresh, boldly printed on its top lid. Other pertinent consumer information such as graded size, factory or farm location, and maybe even a freshness date lets the consumer know what's inside without even opening the carton.

Indeed, in the packaged world, the consumer surrenders the ability to see or touch the product, placing confidence and trust in the package and the purveyor. Of course, in the case of the fragile egg, it's still a good idea to open the lid and check for cracked shells. And that is why the producer does not seal this package: You are invited to check for damaged goods.

Brand-name recognition and a uniform "look" permit the same reassurance of quality and value to carry over into FarmFresh's other products, which taken as a whole, constitute a "line" of similar merchandise. The once anonymous egg no longer is a stranger. It now can be identified with a whole family of familiar products, even though each member is packaged in differently shaped containers

Walter P. Paepcke founded Container Corporation of America in 1926. His interest in modern art and progressive graphic design led to the creation of the innovative ad campaign: Paperboard Goes to War. "Come and Get It—Out of Paperboard" is the headline on an ad designed by Herbert Matter. Man Ray created Can Paperboard Stop a Shell? Paepcke commissioned many internationally known fine artists for company advertising. Fernand Léger illustrated France Reborn in 1945.

fashioned from materials as dissimilar as paperboard and polyurethane. The FarmFresh line of packages all contain, protect, identify and sell their individual food products in a way that the lone egg does not do.

Dissecting the "hows" and "whys" of consumer buying habits is not a new operation. The "art" of targeting and luring the customer had been thoroughly researched and published in books and company manuals as far back as the early decades of the 1900s. Many early works on motivational research were obviously influenced by the heavy hand of psychoanalysis, which had recently become popular.

Marketing theorists Richard B. Franken and Carroll B. Larabee coauthored an amazingly prescient book in 1928 that reads as though it could have been written today.

The preface to *Packages That Sell* states that "Nothing has contributed more to modern methods of distribution than the humble package. Its struggles and development have made marketing history. Today we could not well, and probably would not, get along without it. It has come to be what it is because it fulfills an economic and psychological need. It spells cleanliness, quality, quantity, reliability, discrimination, protection, and, in general, manufacturing and merchandising contentment."

Franken and Larabee's book is definitely an upbeat exploration of how packaging has advanced civilization, helping business and the consumer alike. Packaging controls unnecessary product spoilage. It is a time saver, now that the store clerk is freed from measuring out or weighing unwrapped bulk goods such as salt, flour, or candy. Packaging also stresses trademark and brand names, helping the manufacturer build product loyalty.

A symbiotic relationship exists between packaging and advertising. Advertising provides knowledge of, and creates desire for, a product. In ads or commercials, rarely is the packaged product not shown. But this link was not always so strong. Franken and Larabee reported that only 7 percent of ads seen in magazines in 1900 displayed any picture of the package. Copy-heavy testimonials and decorative borders were the rule, not the exception, for print ads of the time.

It's the soap in DUZ that does it !

DUZ

DOES EVERYTHING

Be thrifty . . . get the giant economy size

Today, the importance of reinforcing the ad message with the package, and vice versa, is a marketing given. The package visually reminds the consumer of the ad, partly through supportive copy: "Perk up with Pep," "A bright wax shine," "A dazzling clean wash." Of course, the package makes the sale. Closes the deal. As long as the product occupies shelf space in the home pantry or medicine cabinet, it will continue to advertise.

Following World War II, the supermarket, a uniquely American institution, became the center for thousands of packaged items.

Packaging, in fact, helped to build this modern-day temple to consumerism. Without the package the supermarket would collapse. The supermarket is a truly modern phenomenon, a logical, impersonal, time-saving system. Specialized stores—the fruit stand, the bakery, the butcher shop, the liquor store— are conveniently located under one roof. Self-service and cash-and-carry have replaced the old-fashioned inefficiencies of credit and bartering. The informed shopkeeper has been nudged into retirement by the "little salesmen" hawking their own wares from box to box and bottle to bottle. They crowd the shelves and cajole you for a moment of your time, a few seconds out of your busy schedule. Packaging stands on its own, to be scanned, critically examined, or ignored, eliminating the need to stand in line to be waited on. Packages can be handled, labels read, and prices considered—all at the leisured stroll or hurried canter of the shopper.

Supermarkets did not sprout from suburban parking lots in the 1950s, as is commonly assumed. An early predecessor was the A&P chain of stores that opened its doors in 1913. One person

Women in fifties and sixties ads and commercials used elegant, sweeping hand gestures drawing special attention to the practical and the mundane (left).

Sometimes women were posed, looking in adoring wonder beside an appliance. The Westinghouse brochure (below) diagrammed the care you needed to give your new freezer so that you, too, could be aglow and seeing stars of satisfaction, c. 1957.

was usually employed to handle a room with an inventory of less than a thousand items. Grand Union was another pioneer. These early economy stores invested as little as possible in hired help and rental property. And it showed. The stores were small and cramped.

The first Piggly Wiggly store followed in 1916. Clarence Saunders, a Memphis grocer, continued to develop the self-service, one-stop-shopping model, and added more merchandise. He also reconfigured the aisles in the shape of a continuous zigzag pattern. Once you entered the store and passed the turnstile, you became the proverbial scientist's rat in a consumer's maze. Shoppers passed through every aisle and walked by thousands of packaged goods until they came to the cash register, where they paid and exited.

In the hard economic times of the Great Depression, selling merchandise required slashing prices. Lower prices demanded a higher sales volume, which in turn called for larger stores. The idea for a "warehouse grocery" was the brainchild of Michael Cullen, a New York merchant. By choosing low-rent locations outside of the city centers, providing night hours, and using aggressive advertising tactics, he prospered. He "crowned" himself, changed the first initial of his last name, and called his offspring King Kullen. His first supermarket was in Queens, New York, and although it boasted a then astounding six thousand square feet of selling space, by today's standards it would be judged an average-sized supermarket.

During the supermarket's early evolution, shoppers used hand-held wire and wicker baskets for their packaged purchases. In the late thirties, Sylvan Goldman, an Oklahoma grocery chain owner, invented a folding contraption with wheels, inspired by the folding chair, that held two wire baskets. After the customer paid for his or her purchases, the folding-chair-on-wheels could be collapsed and stored. In 1947 Goldman introduced the Nest Kart, a "wire carriage" closely resembling today's shopping cart. Now nothing prevented consumers from piling their carts high with lots of packaged goods.

Fortune magazine's October 1953 issue sported an illustration of a shopping cart overflowing with packaged items. An article in that issue reported that supermarkets had captured almost 80 percent of grocery store business. The article concluded, "And until consumers stop moving to the suburbs, buying automobiles, and demanding convenience, there is no reason to suppose the

trend will be checked." Supermarket prototypes may have originated in the Depression years, but it took the prosperity of the post-war years and an avalanche of new consumer products to establish their big numbers across the nation. In 1940 there were 6,000 supermarkets. Twenty years later there were more than 180,000.

Fortune magazine reported new statistics on the buying trends of Americans. Specifically, they were spending a lot more on food—in 1941 Americans spent $20 billion for food; in 1953 that amount tripled to $60 billion. But this new working middle class spent their meat and vegetable dollars differently in the 1950s. Food manufacturers, motivated by fierce competition, called on the packaging industry to create all sorts of modern-looking containers for their consumables. Most of the food being purchased was prepackaged, much of it in individual portions, and precooked. American consumers were channeling their extra food dollars into convenience and "built in" food service. Although packaging innovations such as frozen juice concentrates, pressure-packed whipped cream, and squeezable sauce bottles actually date from the thirties, it was during the fifties and sixties that sales of these and other "easy" foods took off.

While many Europeans sneered at the new Yankee trend of preparing dinner from boxes, cans, and plastic bags, in truth it freed up time for housewives and single people alike. Magazines of the times were rife with cartoons ribbing the modern housewife for cooking meals essentially with a can opener and a pot of boiling water. But it was no joke. Jobs were easier for everyone to get after World War II. Thirty percent of the nation's women now worked outside the home.

The shopping cart and packaged goods took to the cover of the October 1953 issue of *Fortune* magazine illustrated by Jerome Snyder (left). The issue's lead feature, "The Fabulous Market for Food," examined the new buying habits of the country's expanding middle class.

Ben Shahn's July 1949 cover for *Fortune* (bottom) broadcast the business world's early interest in television and the sales potential of commercials.

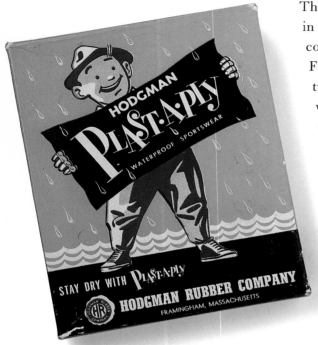

Hodgman Rubber Company of Framingham, Massachusetts, sold waterproof sportswear made of synthetic rubber with the motto: "Stay dry with Plast-A-Ply." The set-up box lid features the company's jaunty mascot decked out in hat and waders, c. 1955.

The baby boom was also in full swing, further complicating matters. From 1947 to 1953, twenty-million babies were born. Between caring for toddlers, holding down a job, and maintaining a house in the suburbs, women found that baking their own bread or cake from scratch was no longer a sign of frugality but of leisure.

Science played a major role in the refining and reapplication of many innovative packaging materials and production processes that had actually lain dormant for a long time. Alexander Parkes of Birmingham, England, developed the first plastic, Parkesine, in the mid-1850s. Produced from nitrocellulose, camphor, and alcohol, it was a thermoplastic material as "hard as horn, but as flexible as leather, capable of being cast or stamped, painted, dyed, or carved."

Ironically, Parkes garnered only a bronze medal for his invention at the 1862 Great Exhibition in South Kensington. Ten years later came the patent for the first plastic-injection molding machine. Yet, oddly enough, another ninety years would pass before this incredibly versatile material was exploited to its full potential.

The use of plastics expanded during World War II. Up until then, they had had limited applications: caps and other types of molded closures, jewelry, and the "skin" for streamlined clocks and other appliances. Lucite also was used by industrial designers and for point-of-purchase displays (page 72).

During the war, America's leading plastic-producing companies teamed with the military to invent ingenious new applications for their product. Army canteens molded out of ethyl cellulose were lightweight and easier to carry than their metal predecessors. Rigid, shock-resistant plastic containers protected ammunition and machine parts. Drugs and first-aid kits were encased in plastic to protect them from jungle humidity. And when the Japanese commandeered the East Indies' rubber supply, the production of its synthetic counterpart, styrene butadiene, shifted into high gear. The Dow Chemical Company rushed the development of polyvinylidene chloride copolymer film, better know by its brand name, Saran Wrap, to be used as a moisture barrier to package parts for guns and aircraft engines. Although technically not packaging, many of these items gave rise to civilian uses after the war.

A versatile packaging material with a much longer history—the common tin can—shares the grocery shelf with its plastic competitors. Originally called a "tin-plated canister," a tin can is actually made from sheets of steel that are tin-plated. Exposed to air or water, uncoated steel will rust. In the early 1800s, the manufacturing of cans was a small-scaled business. Cans were hand-soldered from separate pieces of tinplate, filled, and sealed at the same location. Unavoidably, lead solder dripped into the can's contents in the sealing process. The informed housewife of the time knew to remove any bits of lead that may have fallen into the canned veal she had purchased before cooking up her stew.

Cobb Preserving Company ran the first fully automated canning line in 1898. Two of the largest canning companies in the world, American Can Company and Continental Can Company, were founded a few years later. Beer, a product that had always been bottled in glass or ceramic, was canned in 1935. It first appeared in a cone-top can to accommodate old bottle-filling equipment. In 1959 the Adolph Coors Company introduced the revolutionary, all-aluminum beer can.

Wartime shortages and needs also brought about creative refinements in the canning industry. When America declared war on Japan in 1941, it lost the steady supply of tin it had been importing from the Southeast Asian region now known as Malaysia, which had been wrested from the British by Japan. Because steel was also in short supply, a redesign of the ubiquitous tin can was needed to reduce its thickness without compromising its strength or durability. The solution eventually reached was to roll steel sheets thinner and to reduce the amount of tin through improved methods of electrolysis.

Another invention first used during the war fast became an inseparable part of our "packaging lifestyle." The aerosol (aero + solution) was produced in 1942 by the Bridgeport Brass Company for the United States Army and Navy as a spray insecticide in the Pacific campaign. Nicknamed the "bug bomb," the high-

This Soaky figure is based on the television cartoon character, Top Cat. The show aired on Saturday morning in the early sixties and starred the sophisticated feline, wearing a top hat and vest. The figure's head, manufactured from rigid plastic, unscrews from a squeezable, soft plastic body. Top Cat Soaky rests on a paperboard pedestal with panel copy informing Mom and Dad that they get twenty-five fun baths from the eleven ounces of bubbly liquid. This cat container had a second life as a toy once all the product was gone. Made by Colgate-Palmolive Company of New York, c. 1962.

Three red volts of static electricity shoot from a record on a Stati-Clean aerosol can, a graphic warning of the need for this "automatic" anti-static spray. The type-filled circles of white and red reference the record's graphic shape, c. 1955.

pressure canister with a screw-threaded spray nozzle on top was quite heavy and formidable-looking. A metal valve could be screwed upward for continuous spraying, or threaded quickly up and down for short bursts of liquid poison. After the war, the "bug bomb" was reengineered into an innocuous, lightweight can with a plastic, low-pressure valve inset into a top dome. The first commercial product to take advantage of the new, consumer-friendly redesign was perfume in 1946.

Further control mechanisms for aerosols soon arrived in the form of different types of valves: single shot, continuous release, or metered. In the convenience-crazy fifties and sixties, consumer excitement was high as they speculated what might be marketed next in this ultra-modern type of packaging. The new packaging was expensive, but it was also convenient and easy to use. It seemed as though an endless assortment of products could be propelled more efficiently by the aerosol. By the mid fifties aerosols were on supermarket and drugstore shelves containing hair lacquer (hair spray), paint, shaving cream, furniture polish, air freshener, cosmetics, record cleaner, whipped cream, cooking oil, meat sauce, peanut butter,

pancake mix, salad dressing, butter, cheese, catsup, liquid sugar, cocktail mixes, and, of course, insecticide. Some caught on with the buying public, but many more fizzled.

Developments in twentieth-century art decisively influenced packaging design. Europe was the nursery and hothouse for most the the hybred art movements of the first half of the century. America was playing a game of aesthetic catch-up during this period and first took to modern design like it to took to modern art—very reluctantly. It was disliked on many levels, one being that it was an unwelcome foreign import and a threat to American culture. Eventually, however, this country's middle and upper classes got hip to both modern art and design. Artistic movements such as Cubism and Futurism, along with other modern art movements such as Surrealism, Dada, de Stijl, and Constructivism, would have a lasting impact on graphic design, typography, commercial photography, and advertising, as well as packaging design.

The surrealism of the twenties and thirties became the darling of commerce in the late forties and early fifties. Indeed, Giorgio de Chirico's and Salvador Dali's dreamscape paintings have been conceptually stripmined by graphic and advertising designers into contemporary times. Dali designed packaging for perfume manufacturers using his distinctive style of surrealism (page 78). Paul Klee, Joan Miró, and Jean Arp attempted a more playful surrealist vocabulary. Whimsy and childlike simplicity dovetailed with humor. Miró painted and Arp sculpted curvilinear, amoeba-like forms throughout the thirties. These biomorphic shapes were co-opted in the forties and fifties to express all things modern, be it a package design, a pattern of formica, or the glass top of a coffee table.

When the Nazis officially closed the *Hochschule fur Gestaltung*—the Bauhaus—they unwittingly dispersed throughout the world, and especially the United States, some of the greatest talent ever assembled in one place. Although the early Bauhaus had dedicated itself mainly to the assimilation of fine artist and craftsman, it soon evolved away from personal expression and handicrafts and toward the aesthetics of the machine. Mysticism was out. Rationalism was the new modern religion.

Influenced by de Stijl and Constructivism, the Bauhaus synthesized modern art and design movements. Under the inspired direction of its masters—Walter Gropius, Paul Klee, Theo van Doesburg, Piet Mondrian, Laszlo Moholy-Nagy, Josef Albers, Marcel Breuer, Wasily Kandinsky, Herbert Bayer, and Mies van der Rohe—the Bauhaus created its own techniques to solve aesthetic problems. Typography, product design, furniture, and architectural plans were sold to manufacturers and builders for production. Magazines and books disseminated Bauhaus philosophy and ideas. It was

"Every Package a Show Piece" proclaims this mid-forties trade ad that generously borrows surrealist motifs from the Italian painter, Giorgio de Chirico.

Every Package a Show Piece . . . .

This metal container from Germany held either water or fuel depending on your camping needs. The turquoise label forms a biomorphic lake on which type and illustrations float, c. 1950.

Like a book of matches, Sight Savers tissues are folded and stapled inside an outer cover. The yellow, amoeba-like shape and the isolated, armless hand were common fifties motifs. Dow Corning's packet of glasses-cleaning tissues was a give-away at lounges and restaurants in the early fifties.

Mies van der Rohe who encapsulated the modern design credo with his statement, "less is more." This modern look based on a functionalist, nondecorative aesthetic continues to be a dominant component of contemporary corporate design.

In post-war America, companies that once had cold-shouldered modern design techniques pioneered by the Bauhaus and other European institutions embraced the "new" rationalist approach to design. The "International Style," as some called it, fit like a well-tailored suit for the multinational corporations that had begun to surface in the emerging global marketplace. New York, Chicago, San Francisco, London, Zurich, and Basel became important design centers. Some of the major package design offices that set up shop in these cities were Lippincott and Margulies (New York), Robert Sydney

Dickens (Chicago), Walter Landor (San Francisco), THM Partners (London), Adolf Witz (Zurich), and Verband Schweiz, Konsumvereine (Basel).

Raymond Loewy and Donald Deskey, industrial designers who attained international celebrity for their work in the thirties, began designing packaging as a natural extension of selling their "repackaged" appliances and industrial goods. Raymond Loewy Associates had offices in Chicago, New York, London, and Paris. Industrial designers everywhere expanded their studios to include both graphic and structural designers.

Structural designers advise clients on the most appropriate and cost-efficient material for containing, protecting, and selling their product. Those specializing in folding cartons, for instance, are responsible for calculating exact specifications for the creation of dies used in creasing and cutting cartons out of flat sheets of printed paperboard.

Considerations inherent in such a seemingly simple task are many: What type and thickness of paperboard should be used? What type of flaps should the carton have? Should it have slit-locks or simple tuck-in flaps that facilitate opening? Then again, perhaps top and bottom flaps glued to discourage pilferage are the best solution. Would the carton be filled with product automatically by machine or by hand, as is the case with expensive perfumes that have special inserts and platforms fitted into the carton to display as well as protect the vial? Whether their expertise lay in paperboard, metal, glass or plastic, the structural designer engineered the container with craftsmanship and knowledge of current industry standards before turning to the graphic designer.

In those early years, the preproduction testing procedures and buyer surveys long employed by industrial designers to spot both eye-catching design and the possible "product lemon" were applied to packaging as well. Studio owners sought "scientific" measuring instruments to facilitate, and sometimes to validate, design decisions. The ocular and pupil dilation cameras tested eye movement and emotional response. The tachistoscope projected fleeting images on a screen to ascertain visual perception and memory. The anglemeter tested the visibility of package designs seen from oblique angles—important when a customer rounded a supermarket aisle or glanced at a package at knee level. Designers were split on whether the new dependence on scientific results, which gave the "statistical consumer" the final say, was a godsend or ultimately sullied the profession, producing a glut of mediocre packaging.

Laszlo Moholy-Nagy, one of the founding members of the Bauhaus, thought market surveys inevitably perpetuated bad design. He believed designers should lead and advance public taste, not pander to it. In an introduction essay in the 1959 book *Packaging*, the designer Will Burtin wrote, "Unfortunately, these statistical means of finding out 'what people really want,' 'what makes them buy,' and 'how can they be made to want the product' often result in verbal, printed, and visual assaults on the consumer. More often than not, the findings in these research endeavors seem to confirm a

To ready America for the challenge of Soviet rocket technology, scores of quasi-science kits and toy laboratories were marketed to parents in the late fifties and early sixties to inspire the nation's future nuclear scientists and astronauts. Remco's Science Kit (top) is dated 1961.

Once considered socially taboo, alcoholic beverage purchases by women showed a dramatic increase in the fifties and sixties. To attract this new market, packaging for wines and liquors went upscale. The LeJon brandy carton (left) combines expensive embossing with an unusual structural design, c. 1965.

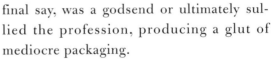

The clowning juggler, illustrated on the lid of Jumbo Tiddledy Winks, works a popular fifties theme—the circus. Polka dot "balls" and the amoeba-like forms of the clown's body were also common motifs.

Cheap dime store toys—a wrist compass and pin—are elevated to space-travel status when viewed on this card illustrated with stars, the planet Saturn, and a space ship. Manufactured in Japan, Space Navigator was exported to the U.S. in the fifties.

time-worn and rather dangerous argument that the majority of people have 'poor taste' and that, in the case of packaging design, the successful sale of a product has little to do with a well-designed package.

As a result the taste standards of packaging are still almost monotonously low, while the technological standards have in a number of countries, and especially in America, reached an extraordinarily high level." Burtin concluded, "While markets, techniques and materials are essential considerations for the designer, packaging is more than the satisfaction of a client and the requirements of a product. To regard a grocery shelf as an art gallery would be absurd, but it is a good principle to think of packaging design as a demonstration that inventiveness and good taste, while fulfilling all the requirements of commerce, can help produce an aesthetically pleasing environment for living people."

Like advertising, package design in the fifties and sixties was big business, and it was growing bigger and more sophisticated every year. In 1963 packaging had become a $23.5 billion industry, ranking sixth largest in the nation. Unprecedented numbers of package designers were able to latch onto

well-paid upper managerial positions.

Many package designers supported the results of scientific research and market surveys. For them such methods constituted honest and valuable tools for creating designs that were responsive to their clients' needs, and simply stated, sold more of the clients' product. Perhaps the best package designers of this period were able to assimilate statistical information without allowing it to subvert or dull their aesthetics. Designers Paul Rand, Lester Beall, Saul Bass, and Will Burtin, whose knack for melding cutting-edge work and demographic concerns appealed to even the most bottom-dollar clients, left behind a legacy of brilliant package design. In Europe and Japan talented designers include Herbert Leupin, Michel Brand, W. M. de Mayo, Milner Gray, Helmuth Winterberg, Egmont Arens,

Hiroshi Ohchi, Toshiaki Yokota, Taizo Shimada, and Kenichi Suzuki.

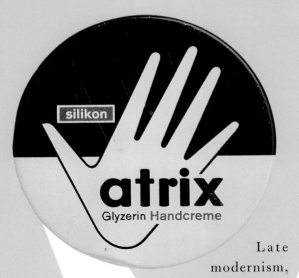

Atrix (left) employs the popular hand motif to good use on this "Handcreme" product from Frankfurt, Germany. Ultra-Kosmetik GmbH manufactured this handsomely designed tin in the fifties.

Improved techniques for printing color photography soon put an army of illustrators out of business. Photography on tins and cartons looked more modern and "real." Bright red lips frame a set of pearly whites on this tin of Pepsodent tooth powder (bottom) containing "irium," Lever Brothers Company's registered trademark for Purified Alkyl Sulfate, c. 1955.

Late modernism, based on early Bauhaus principles of clean lines, efficient use of materials, and form following function, was the "official" philosophy and style of the period from 1945 to 1965. And although it was wildly heralded by designers and industry alike, it was not the only game in town.

Prevailing icons of the period were landing on and sticking to packaged products. When the Soviet Union launched the first artificial satellite, Sputnik, in 1957, the space race was on—and so was middle-class America's love for the "space modern" look. Images of rockets, jets, flying saucers, planets (especially ringed Saturn), the moon, stars, atomic particles, parabolic trajectory curves, and free-form Milky Ways were applied to everything from sewing kits (page 64) to beer cans (page 38). Other popular motifs and patterns used were circus tents, clowns, carnival carousels, scalloped awnings, geometric shapes, ribbons, textural fabrics such as lace and netting, picture frames, polka dots, Victorian etchings, hands, veined leaf outlines, amoeba shapes, tapered forms, Calder-like mobiles, and lots of colorful stripes and bands—horizontal, vertical, and diagonal.

Sans serif fonts dominated much of the graphic design work of the period. On packaging they won out over finer, less bold serif typefaces. Helvetica, a new sans serif font designed by Max Miedinger in 1958, became and is still considered by many the standard of legibility for package design and other graphic media. Other sans serif fonts created in the fifties were Univers, Microgramma, and Optima. Eurostile and Compacta originate from the early sixties as does the funkier font, Adlib, designed by Freeman Craw in 1961. Less conservative packages sported these new fonts in playful hues to communicate a fun, modern product.

Lulled into an unrealistic belief in an ever expanding, always-improving economy and a lifestyle dependent on

homogenous and a lot more grown up and wary. This transitional period would eventually lead design professionals away from purist Bauhaus tenets and toward experiments in "eclectic historicism"—or, as it became coined—Postmodernism.

The packaging for The Kingston Trio retrospective successfully evokes a fifties aesthetic, while maintaining the look of a contemporary CD collection. Designed by Vittorio Costarella (illustrator) and Michael Strassburger at Modern Dog in Seattle, Washington, with art direction by Jeff Fey and Tommy Steele at Capitol Records.

planned product obsolesence, the postwar consumer's joyride skidded to a sudden halt in the mid sixties. Other nations became the young turks of modern industrial efficiency and production. The "happy days" after World War II had also been a time of confining conservativism, racism, and sexism. And creeping fear of the other big "ism"—communism—had created a perceived need for social control and a deep-rooted sense of paranoia.

Acting as a cultural barometer of sorts, graphic design and advertising responded to the societal upheaval of the late sixties. New ways were found to communicate and sell to a nation that seemed far less

Today, Late Modernism, the professional style of the mid forties through the mid sixties, is once again a source of inspiration for contemporary designers. While today's designers often quarry the rock pile of kitsch motifs, they are equally influenced by the more sophisticated work of Saul Bass, Paul Rand, Alvin Lustig, and Alexey Brodovitch.

The packaging line for Groove Dots chocolates, for instance, created by Laurie Rosenwald, lets loosely drawn, Picasso-inspired images doodle across flat, brightly hued fields and coexist with patterns of clouds, stripes, and polka dots. The front panel of John Sayles's package for the Curtis Paper Company is die-cut to show off a toothy grin of crayons, his rough linework

Haley Johnson, and others are reintroducing Late Modernist packaging design to shelves.

The following collection of packaging boasts some outstanding and unusual pieces from the mid-twentieth century. Whether included for their strong graphics, classic fifties motifs, social commentary, or quirky humor, they are all cultural time capsules from a rich and often contradictory era in design.

John Sayles, working out of Des Moines, Iowa, plays with Curtis's new line of colored papers by creating a carton of crayons as a promotion piece.

Morozoff, a venerable, long-time chocolatier in Japan, commissioned New York designer Laurie Rosenwald to create a packaging line for Groove Dots chocolates. Her drawings and hand-rendered typography are beautifully integrated throughout this unusual collection of cartons and boxes.

and the hand-penciled type echoing a fifties' sensibility. In both designers' work, childlike simplicity underscores witty and handsomely executed packaging. Vittorio Costarella and Michael Strassburger at Modern Dog, Neil Powell at Duffy Design,

heat
and serve

American food manufacturers, caught in an increasingly competitive market after World War II, began selling their products in containers that seemed better suited for sputniks than for down-to-earth, Formica-covered kitchenettes. Housewives squirted, squeezed, and sprayed all sorts of unlikely foodstuffs from foil pouches and bottles, aerosol containers, and plastic tubes. Seasonings were packed into capsules like vitamin pills.

Many of the breakthroughs in food processing and packaging were the by-product of experiments for the real and imagined needs of jet and space travel. The commercial airlines pioneered "pouch cooking." Frozen in mylar-polyethylene bags, single servings were cooked in boiling water and then squeezed onto the plates of hungry travelers. Even an untrained steward could now turn out a complete meal with less dexterity than it took to scramble an egg.

With children to care for, large tract homes to clean, and maids in short supply in suburbia, the average middle-class housewife relied more and more on the new packaged foods that promised greater convenience in preparation and serving. Instead of loading the shopping cart with all the bulky ingredients needed to bake a chiffon cake, housewives could simply buy a box of Duncan Hines's cake mix. Compact, lightweight, and easy to make, box mixes and other food packaging with "built-in maid service" also saved cabinet and refrigerator space.

In 1954 Swanson introduced the TV dinner—the Cadillac of frozen fare. Packaged in a carton designed to look like the front of a television console (complete with wood grain and control knobs), it contained an entire dinner in an oven-proof aluminum tray. On screen the package displayed a festive dinner swimming in melted butter and gravy. What supermarket shopper could resist the TV dinner's novelty and handiness? No pots to clean, no chopping or measuring to do, no dishes to wash and dry. Life was never so easy.

TV dinners became an overnight hit, especially with kids. The unique aluminum tray, however, would become a casualty of the microwave age. One of the originals is safely enshrined in the Smithsonian Institution's collection.

Express Dairy Co. Ltd. of London sold individual servings of tuberculin tested milk in disposable, triangular-shaped pouches. Snipping off a corner opened the heat-sealed package for easy pouring or for inserting a straw—a rather progressive idea for 1958.

One of the most over-worked words on fifties packaging (along with "new" and "convenient") was "instant." New, Instant Pream was a formula of powdered cream, skim milk, and lactose. Made by M & R Dietetic Laboratories, Inc. of Columbus, Ohio, Instant Pream conveniently needed no refrigeration, c. 1958.

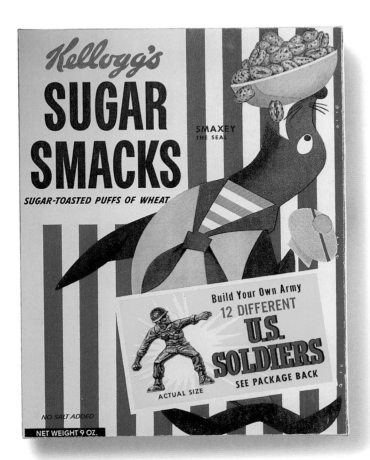

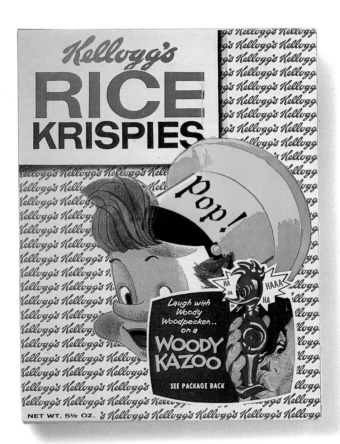

Although the fifties decade is often credited with the introduction of ready-to-eat, just-add-milk cereals, familiar brands such as Kellogg's Rice Krispies were first launched in the twenties. These two cereal cartons, however, display classic fifties graphics with their cartoon characters, premium offers, and references to television. In later years Sugar Smacks was renamed Honey Smacks. Sugar Smacks and Rice Krispies cartons dated 1957 and 1958, respectively.

"Save these bags" extols copy on the Howdy Doody fudge wrapper. Airing in the early fifties, the first popular children's television show starred a cowboy marionette and his human costar, Buffalo Bob Smith. Its companion ice cream bar wrapper from the same period—"Sure Good"— says it all.

Bing Crosby, a well-liked singer, actor, and comedian in the forties and fifties, licensed his own brand of ice cream in 1953 with a pledge of quality printed on its top panel. It assured the buyer that the product was processed in compliance with the rigid specifications "as set forth in our 'Cream of the Stars' quality formulae." Valley Farm's Bing Crosby Ice Cream was manufactured in—where else— Hollywood, California.

Made by The Borden Company in New York, this ice cream carton from the mid fifties provided a handy tear strip for easy opening. The popular "Elsie-Daisy" trademark was designed by Frank Gianninoto in 1952.

"Oh-Gee!" yells the blue-eyed, all-American boy as he chomps on a candy bar the size of a brick on the cover of Big Time bars. This colorful carton containing twenty-four big ones was distributed by Hollywood Brands, Inc. from sunny Centralia, Illinois, c. 1955.

Barton's Bonbonniere tin of Almond Kisses sells hip sophistication with its lively scenes of Parisian street life. This confectionery from Brooklyn, New York, had sixty-five owner-operated shops located in major U.S. cities in the mid fifties.

Created in Paris by Kréma for the American market, this fanciful, miniature kiosk once contained assorted candies. Diminutive posters, showing popular French entertainers of the forties and fifties—Edith Piaf, Maurice Chevalier, and Marcel Marceau—circle the center fiberboard tube. The hand-painted, finial-shaped roof is a wood stopper, c. 1950.

Spartan Stores of Grand Rapids, Michigan, marketed their own brand of packaged spices. With name brand emblazoned overhead, the tin of sage designed in the early fifties (top) displays a battle-ready Spartan. Spartan's alum tin from the mid sixties (bottom) sports a toned down and simplified redesign. A pair of Grecian-like borders have replaced the company's sword and shield toting namesake. The color bands are flip-flopped on the sixties container with top shelf going to the spice this time, not the brand name.

RED
OWL

NET 1½ OZ.

APPLE PIE
SPICE

A BALANCED BLEND COMPOSED
OF CINNAMON, MACE AND CLOVES

RED OWL STORES, INC., MINNEAPOLIS, MINN.
DISTRIBUTORS

ANN
PAGE

PURE GROUND

Sage

Ⓤ   NET WT. 1 OZ. 28.35 G.

NET WEIGHT 1 OZ.

ZEST
FLAVOR IMPROVER

99⁺% MONOSODIUM
GLUTAMATE

A. E. STALEY MFG. CO. • DECATUR, ILL. U. S. A.

Who could say "no" to a salad like this?

Why, even the hardest-to-please eaters in your family will cheer when you bring this Jell-O salad to the table!

And its shimmering good looks and whole-family appeal aren't its only advantages. You can make a Jell-O salad hours—even a full day—ahead of time, and serve it in all its glory when you get good and ready to!

Why not serve a Jell-O salad *tonight?* It's one dish *everyone* will want!

Now's the time for JELL-O SALADS!

Accident-prone nursery-rhyme characters Jack and Jill spy a formidable hill of slippery gelatin to climb on their namesake carton of "the aristocrat of all gelatin desserts." Manufactured by Junior Food Products Company, Inc. of Tyrone, Pennsylvania, back panel copy boasted that the "modern development of new, perfect inner wrap prevents loss of flavor," c. 1955.

Jell-O has been America's favorite gelatin dessert for over a hundred years. In this 1952 ad, a rather unusual recipe for "garden salad ring" has the whole family shouting for joy as Mom is carried victoriously to the table. Jell-O is a registered trademark of General Foods Corporation.

Discovered! *The secret of making the ♕ueenly Chiffon from a mix!*

New Duncan Hines
Chiffon Cake Mix

*Now you can make this
fabulous ♕ueen of Cakes in minutes!*

Designed with a homey, checkered-tablecloth pattern, the Duncan Hines's cake mix package (lower right corner) in this 1958 ad contained all the ingredients to bake this regal-looking chiffon beauty. Best of all, you could create "this fabulous Queen of Cakes in minutes!"

At the intersection of Milwaukee and Devon Avenues on Chicago's northwest side, Superdawg is a shrine to midcentury America's two great shaping influences: the automobile and mass consumerism. Maurie Berman created a hot dog stand in 1948 topped off with two giant wiener statues. On the roof, a frankfurter guy flexes his biceps in Tarzan apparel alongside a demur and beribboned frank girl. The design on the containers for Superdawg's drinks, fries, and hot dogs have changed little over the years. The signature characters, diamond pattern, and script logotype with the giant *S* all reference the drive-in's architecture.

Raymond Loewy Associates redesigned Armour Star's extensive line of meat product packaging in the mid fifties. At right is a printed paper label for a can of Armour hash.

Two Rath Black Hawk ham tins are illustrated in the bottom right corner of this point-of-purchase, plastic, vacuum-formed sign. Each package is emblazoned with Rath's stylized warbonnet logo. A fifties-style, cartooned family looks hungrily on at a realistic platter of sliced ham, c.1958.

bottoms up

Beer made the first successful break with glass-bottle containers and into flat and cone-top cans in 1935. Today, the beer can is one of the most enthusiastically collected forms of packaging.

Beer has been bottled and sold in glass and ceramic containers since the mid-nineteenth century. Prior to that time, if you wanted a beer "to go," you had to have a bucket or jug on hand to take the frothy brew home from your local tavern. A Montana brewer canned the world's first beer in 1909 but eventually had to "can" his experiments because of failure. There were two problems. First, although the cans of the time were thicker and heavier, they were not structurally as strong as those of today. Seams would burst under the pressure needed to contain the carbonated brew. And, second, the can's tin plating reacted chemically with the beer. The result was a murky, malodorous liquid.

Over the years, different types of inner beer-can linings were tried. Early forms of enamel and pitch, used to coat the insides of beer kegs, were unsuccessful for reasons of taste, aesthetics, or difficulty of application. In 1934, just one year after Prohibition had ended, the American Can Company patented a special nonpenetrable lining, trademarked "Keglined." Several American brewers sold canned beer the following year.

At first canned beer was sold in flat-top and cone-top containers. Continental Can developed the cone-top version because it didn't require expensive new filling machines at the breweries. Unlike the flat-top, it resembled the bottle's form and could use the old machinery with some modifications. It also could be opened with a familiar bottle opener. The flat-top needed a special can opener, fondly christened the "church key."

The flat-top beer can soon won out over its cone-top rival as larger breweries eventually invested in new filling equipment. Comparatively, the flat-top can was less expensive and was easier to stack and ship. And the cone-top had one slight marketing flaw: it resembled metal-polish containers.

Created by Gettelman Brewing Corp., Milwaukee, Wisconsin, this beer label appealed to modern tastes with its faddish icons: a portable television set, wire furniture, and potted tropical plant. The old logo wreathed in laurel leaves is mirrored by a modern-day Bavarian hoisting a foamy one.

The "six pack" of twelve-ounce beer cans on these two pages exhibits two fifties marketing strategies for selling the frothy brew. The beer can designs on the left page aroused the interest of drinkers who identified with products that had an up-to-date and progressive image.

Jet Near Beer, Genesee Cream Ale, and Bosch Premium Beer have a look that's stylishly modern. One of the first canned beers selling to weight-conscious consumers, ultra-modern Jet was "ideal for low calorie diets."

Wisconsin Club Premium Pilsner Beer, The Master Brew, and Bavarian Club Premium Beer call upon traditional images of Bavarian landscapes, lederhosen-wearing waiters, and heraldic shields to sell to beer drinkers who took comfort in the traditional icons of old-world brewing.

Pacific Brewing and Malting Co. of San Jose, California fashioned this part plaster and part beer-bottle advertising figure for John Wieland's Extra Pale Lager Beer that looks more than a little like fifties television star Jackie Gleason, c. 1950.

Cresta Blanca Wine Company distributed the Coronet VSQ Brandy display to liquor stores in the late forties. The congenial waiter has a head shaped like a brandy snifter.

Draft-brewed Blatz beer from America's brew capital—Milwaukee, Wisconsin—cast hand-painted, metal figures with bottle, can, and beer barrel bodies for store and back bar display, c. 1950.

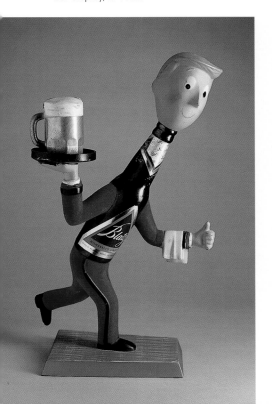

With names such as Pep-Up, Dash, and Pep, it's an easy guess that these soft drinks used sugar as a main ingredient. Meant to be returned and reused by the manufacturer, the three bottles have baked on labels (commonly referred to as a "painted label") that made washing and refilling more efficient than paper labels. Typically with labels screened on green glass in white and red, the seven-ounce bottle was the container of choice, although Pep holds a "King Size" ten fluid ounces, c. 1950.

Perrier bottled carbonated water of France ran this ad reprinted from a poster by Jean Carlu. A big-eared clown listens to "the water that goes fizz. . ." Perrier's signature green glass bottle and label designs have changed little over the years since this ad, c. 1946.

Illustrated by Raymond Savignac, one of the greatest poster artists of the twentieth century, this humorous French advertising poster from 1949 for Pils Ale graphically shows how you'll double your drinking pleasure.

Wynola and Rummy easel-backs are from the forties and fifties respectively.

A whimsical, cap-and-gowned "Sir Cola-Nut" urges you to "graduate to Wynola." Copy on the bottle's backside reveals that the transparent glass is "for your protection clearly revealing the purity of its contents—see what you drink!" Wynola was a product of Bell Bottling Co. Limited of Winnipeg and Manitoba, Canada.

To avoid the obvious rum concern or expectation, the front of Rummy's lithographed bottle clearly spelled out: non-alcoholic. Two darkened faces with bodiless heads invite you to "get chummy with Rummy."

These miniature bottles are salt and pepper shakers in disguise. The Squirt Company manufactured the premium to promote its "exciting new kind of soft drink. . .with fresh fruit flavor you can see!" Side panel storyboards show where the fun, novelty shakers could be used, c. 1955.

Throughout the twentieth century, advertisers played on American insecurities about bodily embarrassments and perceived health risks. In the thirties, magazine ads warned of the damaging effects of constipation. Forties advertising blamed loneliness and the lack of dance partners on body odor. With incomes on the rise and the middle class expanding, many Americans in the fifties were indulging in richer, high-caloric foods and snacks. Thus, the American obsession for "diet" drinks and foods was cultivated.

Graft's sugar-free root beer sold to the "figure wise." The thoroughly modern, wasp-waisted woman, illustrated on the can, advertised the benefits derived from consuming a soft drink artificially sweetened with calcium cyclamate, c. 1950.

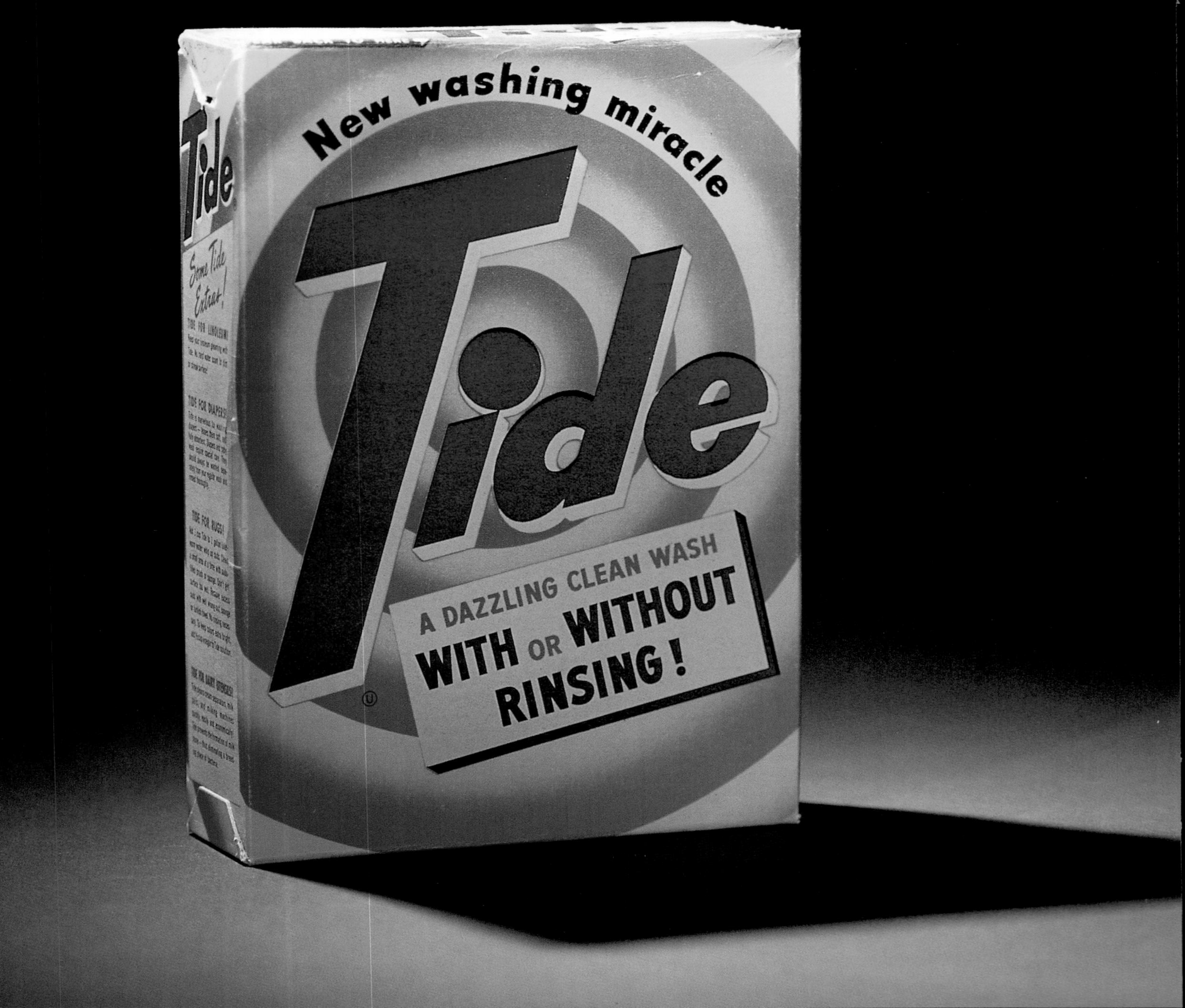
Tide
Some Tide Extras!
New washing miracle
Tide
A DAZZLING CLEAN WASH
WITH OR WITHOUT
RINSING!

clean
and
shine

Launched in the U.S. in the late forties, Tide targeted heavily soiled laundry with a new type of synthetic cleaning base. After fifty years, it continues to be one of the top selling detergents in the country. Although the same design can still be found relatively unchanged on American supermarket shelves, in Britain the package was completely redesigned in the mid sixties.

As in the fifties and sixties, today's detergents get your clothes' whites whiter than white through chemical cleaning agents. Prior to World War II, however, soaps were largely made from animal fats and lye. Companies that processed fats often produced both shortening and soap—one product to cook with and one to clean with.

Shortages of fats and natural oils in Europe during World War I spurred research in soapless detergents. But it wasn't until 1947 that a major American company got serious about marketing a cleaning product based on synthetic compounds. That was the year that Procter & Gamble introduced Tide to Americans, who had previously been washing their clothes on a high-cholesterol diet.

Created by Donald Deskey, famed thirties industrial designer, with the help of marketing psychologist Louis Cheskin, Tide's packaging is a textbook case of the new aggressive marketing methodology of the forties, fifties, and sixties. The concentric circles of orange and

yellow dominate the box's front panel and suggest heat, power, and strength. If this were food packaging, the colors would be only appropriate for spicy foods. For clothes-detergent packaging, heavy-duty cleaning is the message communicated. A three-dimensional, sans-serif "Tide" floats above the swirling agitation— white sides with a top veneer of cool indigo blue. The huge T in Tide is positioned like a no-nonsense tool, ready and waiting to hammer out dirt and stains.

In comparison, Procter & Gamble's Dreft package (page 51) spoke in soft pink and green hues to the customer who wanted a mild detergent to wash the new delicate fabrics made of nylon, Orlon, and Dacron. Typically, light colors on packages, if deployed over a large area, make the box look larger. Dark-color backgrounds make the box appear smaller and heavier. Spelled out in a friendly, sans-serif, all-lower-case font, this brand name announced from the shelf that here was the detergent to buy when gentle cleaning action was needed.

Prompted by a U. S. Senate subcommittee investigating deceptive packaging, new legislation passed in the mid sixties regulated net weight and ingredient declaration. Prior to the new law, boxes and cans net-content statements were seldom "clearly, prominently and conspicuously shown on the main display panel."

The four boxes filled with clothes detergent and soap (Procter & Gamble's 1950 Duz formulation was a fat-based soap and not a synthetic detergent) on these two pages are approximately the same size, but are labeled differently: regular, large, and giant. Try to find the declared weight on any of these fifties boxes.

Shina Dish, manufactured by Tidy Products Co. of Omaha, Nebraska, is designed with the same "loud" lettering and colors as clothes detergents. The front and back panels carry several quick-read messages: "Now. . .for dishes and fine fabrics!," "Really does save your hands," and "Dishes shine, glassware sparkles—without wiping!," c. 1951.

Brillo's soap pads, boxed in their distinctive red, white, and black packaging, made "scorchy" aluminum pots and pans look like new according to ads in the forties and fifties. Former graphic designer Andy Warhol turned the package into modern art and a pop icon by silk-screening it on large plywood cubes.

Body Sheen claimed its thick-liquid cleaning and polishing agents left your car dazzling without a wash job. The product's alternating letter colors set against a white background lent a lively but clean look, c. 1958.

Johnson's Holiday car washing cream sold in a convenient soft plastic squeeze bottle. The multi-colored "holiday" type carousels up and down in typical, fifties-style playfulness. Manufactured by S.C. Johnson & Son, Inc., Racine, Wisconsin, about the same time as the Body Sheen tin.

A company known for its car products since the thirties, Plastone of Chicago, Illinois, produced this pair of Turtle Wax products. The tin of car wash with plastic screw cap (right) sports a dapper-looking turtle mascot outfitted in ascot, diamond stick pin, monocle, and top hat, c. 1955. The green glass bottle with metal cap (left) shows off the handsome redesign executed a few years later. The retooling is distinctly modern. Gone is the extraneous copy and detailed illustration. The turtle has evolved into a stylized logo, simplified, yet full of personality and charm. And still wearing his chapeau.

The three polishing cleansers packed in fiber-board tubes with metal end pieces date from the mid forties to the early sixties. Their company trademarks, however, span a much broader time period. Kleen King's rotund monarch and Air Maid's wand-wielding fairy first appeared in the mid forties and mid fifties, respectively. Louis H. Soule created Bon Ami's "hatched chick" in 1901.

Esquire's Scuff-kote carton from the fifties (left) appealed to kids with its circus-themed graphics. The handy applicator and "tip-proof" bottle made moms happy, too. The Scuff-kote packaging for Super White (right), especially formulated for children's white saddle shoes, is an early-sixties redesign.

This special promotion from the Purex Corporation, Ltd. of South Gate, California, appears to be a "buy one/get one free" offer. Smaller print informs the consumer that you only get one can at half price, if you buy one at the regular cost. In 1956, New Blue Dutch Cleanser modernized its character mark from a Dutch woman chasing a goose with a stick—"Old Dutch cleanser chases dirt"—to a woman with a wand but no goose. New Blue was now "the modern cleanser for modern home-makers."

The unopened bottle of Glamorene Foam Upholstery Cleaner with accompanying carton features a band of heat-sealing plastic tape wrapped around the bottle's glass neck and the red metal cap to prevent any product leakage. Three graphically rendered pieces of furniture on the box and the bottle's label visually inform the consumer that Glamorene will work its foamy wonders on your Sears sofa as well as that modern, armless chair from Herman Miller, c. 1953.

Hard Gloss Glo-Coat, marketed for use on all floors including asphalt tile, rubber, vinyl plastic, wood, and terrazzo, was a product of Johnson's Wax, which headquartered in the Frank Lloyd Wright designed office building in Racine, Wisconsin, c. 1955.

around
the house

A benchmark of the fifties, 1954 witnessed cultural and technological precedents that set the tone for American consumerism for the next ten years. In the summer of 1954, Elvis Presley's radio debut started a sensation that continues to affect American popular music. The song aired: "That's All Right, Mama." On Thanksgiving Day in the same year, the first TV dinner—turkey with mashed potatoes, gravy, and stuffing—sold in supermarkets across the country. And in December, the world's first pocket-sized, transistor radios appeared in department stores. If you were the lucky receiver of the $50 Regency TR-1 for your Christmas or Hanukkah gift, you had the hands-down envy of every kid on your suburban block.

Invented in the late forties at Bell Laboratories, the transistor ushered in the age of microelectronics. Before transistors, radio and television sets were crowded with bulky vacuum tubes and a tangle of criss-crossing wires. After initial high prices, transistor radios soon became affordable.

With the new age of miniaturization, people no longer had to accommodate themselves to the product. Now the product began to fit the lifestyles of its buyers. Once families had been compelled to gather together around the radio, which was a heavy piece of furniture anchored in the living room. In the fifties and sixties, portable radios and televisions could be enjoyed at the beach or in the privacy of one's own bedroom, a handy item for teenagers.

The Regency transistor radio won an award for product excellence from the Industrial Design Society of New York and was included in the 1955 American Art and Design Exhibition in Paris. Belying the modern, advanced technology of the product inside, the Regency's package looked more like men's after-shave cologne with its spare design. Japanese and Hong Kong manufacturers, who later dominated the production of transistor radios, also largely ignored the value of good package design. Paradoxically, and with rare exception, this icon of fifties and sixties modern living—the transistor radio—was marketed with less than stellar packaging.

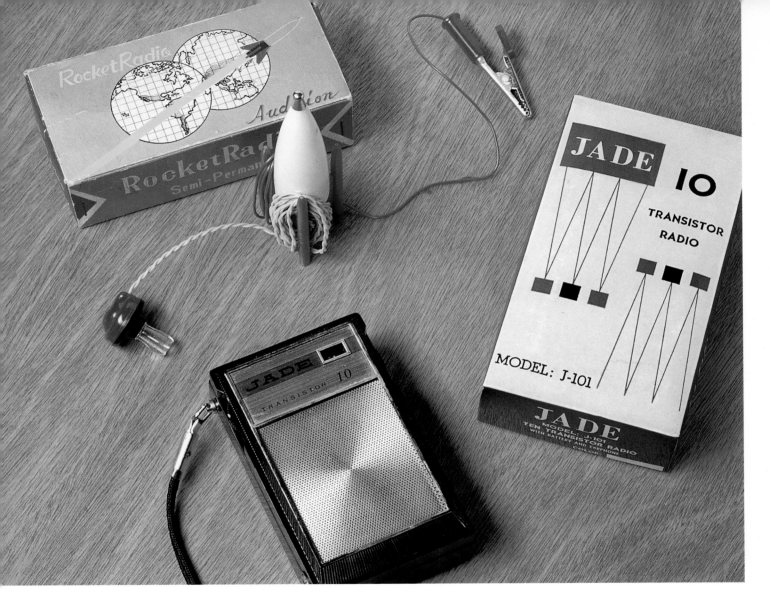

RocketRadio Aud-ion, manufactured in Japan, had only two transistors (the usual number was six) and a simple tuning method: Just pull out or push in the rod. This rocket sold in earth stores in a crudely designed package, c. 1959.

Jade Model: J-101, made in Hong Kong, carried a hefty ten transistors and exhibited more sophisticated package design. Printed in two colors, green and black, its graphic lines and squares symbolize with modern elegance wires connecting to miniature transistors, c. 1964.

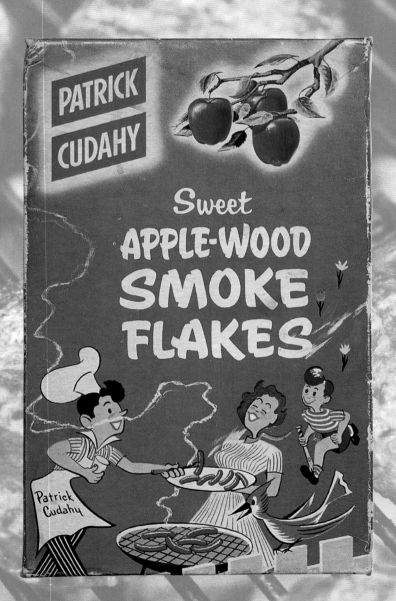

More than any other American family activity, backyard barbecuing symbolized the good life. After World War II, the suburbs became the home for white, middle-class couples wanting to leave the racially and ethnically diverse cities. On the front and side panels of Sweet Apple-Wood Smoke Flakes, the whole family comes running when they get a whiff of the irresistible aroma of franks on the grill. Produced by Patrick Cudahy, Inc. of Cudahy, Wisconsin, c. 1954.

These sixties matchbox labels are colorful, miniposters from the Netherlands, the Scandinavian countries, and Germany.

What does space travel have to do with darning needles? Nothing. As Japan recovered economically from its loss in World War II, manufacturers cashed in on the American space craze to sell their products. These assorted needle envelopes date from the early fifties.

64

Two scenes of domestic family life decorate these needle holders from fifties Japan. Reliance needle book targets the thoroughly modern. Dad is welcomed home by his loving family after a business trip requiring air travel. Sewing Susan underscores societal reinforcement of a woman's domestic duties.

Specializing in insecticides containing DDT, the Swiss company, Geigy, employed an overall-wearing elephant as its mascot. This metal pump spray was manufactured for the Spanish-speaking market, c. 1955.

Both by Bugs

Mow 'em down instantly!

The Mabex Company of Philadelphia, Pennsylvania, distributed this carton of moth flakes containing 100 percent refined naphthalene. A graphically rendered moth does a tailspin against photographed woolen fabrics, c. 1960.

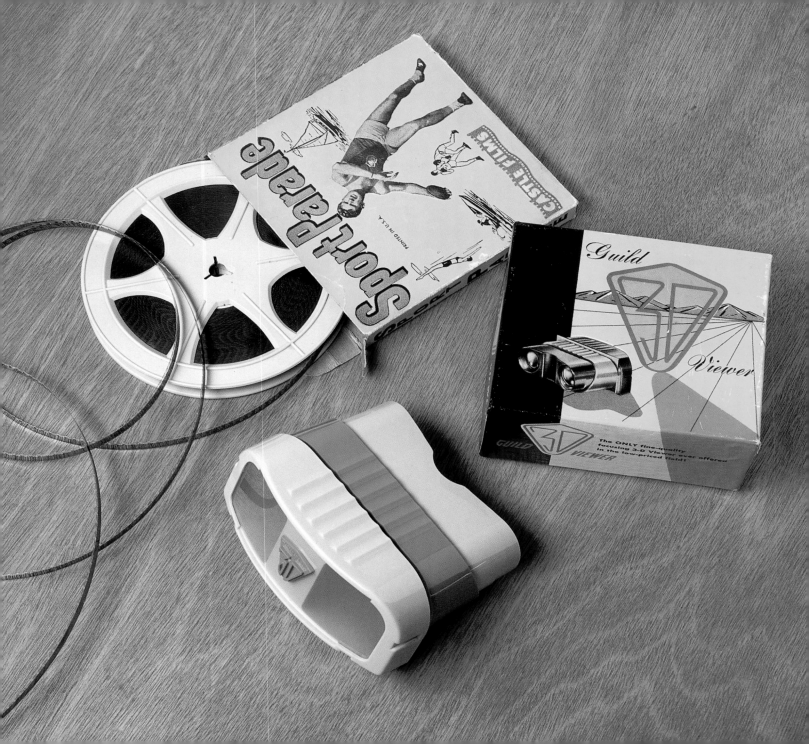

Castle Films released 8mm captioned silent films for home viewing in the forties and fifties for both child and adult audiences. Part of the Sport Parade series, this carton contains a reel titled: "Swimming and Diving Aces."

Manufactured by Craftsmen's Guild of Hollywood, California, the attractively designed, plastic 3-D Viewer sold in a box illustrated with subtle surrealist motifs: long cast shadows and a flat, perspective-lined field that leads the eye to an anonymous mountain range, c. 1948.

The bottom tin of Rema patches from the late forties has a stylized Autobahn running through a giant tire lithographed on its lid. The sixties redesign on top gains in sales copy and illustrated product uses, but loses aesthetically. Both Rema self-vulcanizing rubber patch tins were made in West Germany.

Lester Beall designed Simoniz' line of cleaners and polishes in 1951. The bold *S* provides a strong graphic image that reinforces brand name recognition. The Simoniz Company began operations in Chicago, Illinois.

Sinclair Refining Company, headquartered in New York, ran a series of advertisements in 1930 using dinosaurs to visually connect the idea that the oldest crude oil made the best lubricant. In 1931, a brontosaurus, Dino, gained exclusivity over his brethren as Sinclair's charactermark. These two tins show off the strong packaging and brand identity that the refining company possessed in the late fifties.

Two forties and fifties companies, Hastings Manufacturing and Mac's Super Gloss, used the same three colors—red, yellow, and black—to package their separate lines of automotive products. While Mac's used the "thrifty Scotsman" as their character trademark, Hasting's brand of Casite paraded the unlikely choice of an effeminate jailbird to symbolize its corporate image.

ladies and gents

Women's cosmetics and perfumes comprise a luxury market that has never suffered for lack of creative and beautiful packaging. The box for a costly perfume provides all the essential needs of an ordinary package, protecting and identifying its contents. But it also does something extra and special. The well-designed perfume box attracts the customer and arouses an urgent desire to own or to purchase it as a gift. It communicates personality, attitude, and exclusivity. Its high perceived value goes with its equally high price tag.

In the forties and fifties, famed Paris dressmaker Elsa Schiaparelli also designed imaginative and whimsical perfume bottles and boxes. Influenced by her good friend, the flamboyant surrealist Salvador Dali, Schiaparelli created packaging for the new scents Zut and Shocking (pages 76 and 78). The bottle for Shocking, by far one of the more exotic flights of surrealist-inspired perfume packaging, was molded in the shape of a Victorian dress dummy equipped with a cloth measuring tape and a tiny bouquet of artificial flowers sprouting from its "head." Made even more precious and incongruous under a miniature glass bell jar, Shocking combined elegance with a touch of the bizarre.

Surrealism, the modern art movement that had its heyday before World War II, gained popularity in the arts again after the war and was a style frequently exploited in graphic design and advertising. Salvador Dali, René Magritte, Marcel Duchamp, and Man Ray were some of the movement's best-known artists.

As in surrealism, body parts used as decorative aspects—the eye ball, the lips, the armless hand, and the bodiless head—shocked and delighted the consumer. Liquid Liptone's point-of-purchase display constructed of lucite and painted plaster works a familiar surrealist vocabulary of motifs: the face-like mask and the hand sans body. The mouth on the plaster carton's panels recalls Man Ray's painting of giant red lips floating in the sky and Dali's famous settee in the shape of Mae West's ample crimson lips.

Manufactured in Chicago, Illinois, Princess Pat was a popular line of cosmetics that retailed in drugstores. This fragile store display for Liquid Liptone, constructed of plaster of Paris and lucite, survives over fifty years after it was created.

73

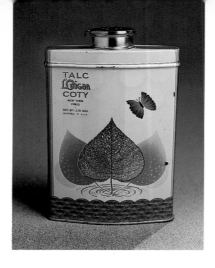

These similar talc tins were fashioned with a European flair for style and subtle design. L'Origan's delicate leaf and butterfly reflect Japanese-like elegance. L'Aimant displays an overall pattern of pink, cream, and gold faux shark skin. Both tins are by Coty of New York and Paris, c. 1945.

Intoxication's one ounce sachet concentrate by D'Orsay of New York and Paris retailed in a carton illustrated with lively, sexy drawings.

Cachet

from France
Lucien Lelong sends
you his newest
most brilliant
perfume...

LUCIEN LELONG

With Paris as a backdrop, this ad for Lucien Lelong's extravagently packaged Cachet recalls the long tradition of quality French perfumes, c. 1953.

"Zut," is the English equivalent of "Oh, damn!" It's also the name of a perfume by the wonderfully eccentric French dress designer, Elsa Schiaparelli. The names of the most successful perfumes in the forties and fifties were strange and exotic, evoking escape and the breaking of society's stuffy rules. This advertisement for Zut, appearing in the fifties magazine *Flair*, was drawn by Marcel Vertes.

Packaged in a cream-colored box with gold foil stamping, Amour Amour is a statement of confident simplicity. This Parisian perfume was the creation of Jean Patou, c. 1950.

The carton design for Norelco Coquette says stylish and modern with its illustration of a woman dressed to the nines and its smart printing in black, pink, and metallic gold. The rotary electric shaver that was "made in Holland and styled in Paris" came in a zippered, leatherette travel case, c. 1954.

Sentinel Products of Cleveland, Ohio, distributed the Cotton Plucks tin, with its convenient top dispenser, in the late forties. The ubiquitous armless hand, used as a surrealist decoration throughout the forties and fifties, is shown on every panel holding a wad of cotton.

The tailor's dummy shows the surrealist influence in this flacon's design for Shocking by Elsa Schiaparelli (left).

Salvador Dali designed this luxury perfume called Le Roi Soleil. A hand-cut crystal stopper of a sunburst with facial features comprised of birds rests on a crystal base (below left).

Bourjois' Mais Oui and Soir de Paris scented powders sold in drum-shaped boxes in the forties. Silver foils and metallic inks add a touch of understated elegance to these containers made in Mexico (above and left).

The packaging and flacon label for Liberty perfume were created by the English designer Ashley Havinden in the late forties.

This cylindrical box of Contraband dusting powder and perfume vial by Tussy in gay, tantalizing pink and gold sold in New York and Paris markets around 1962.

W. M. de Majo designed By Candle Light perfume's packaging in 1951. A snuffer-cap, pyramidal top slips over the folded sides of the base. All is held into place by a slip-ring cord.

TIGER BRAND

THE CHOICE OF OUT-DOOR MEN

ESTABLISHED 1851

TRADE MARK

REGISTERED

TIGER BRAND

MANUFACTURED BY GALT KNITTING CO. LTD., GALT, ONTARIO

Tiger Brand's easel-back store display, "the choice of out-door men," shows male bonding while selling underwear to men and boys. Dad and Junior cast shadows revealing their true hunting spirits, c. 1948.

Healthknit sold men's underwear with "cantilever action" for the colder fall and winter months in this printed cellophane over-wrap. The illustrated model on the right demonstrates his comfort by lighting up a cigarette, c. 1952.

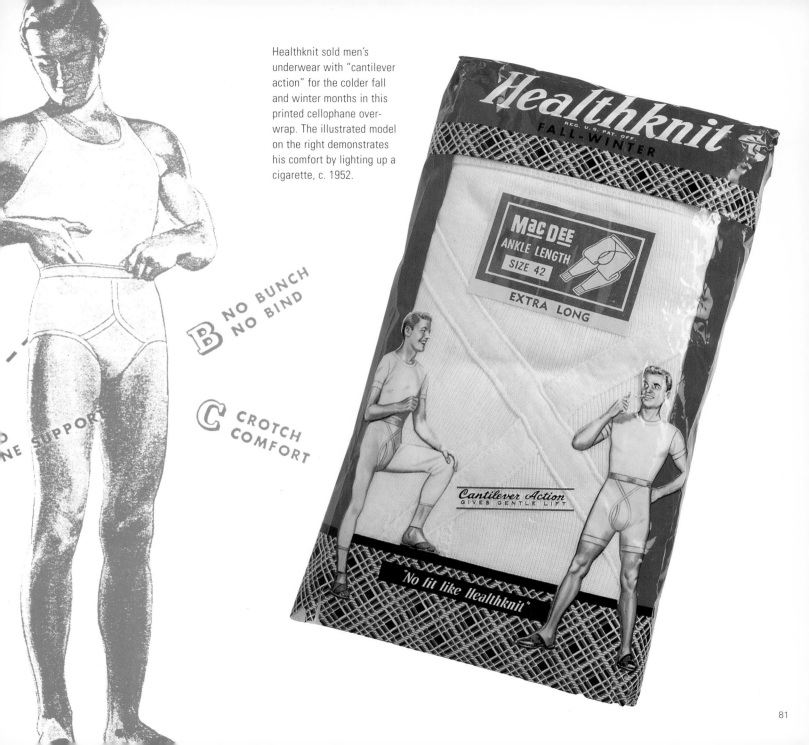

B NO BUNCH
NO BIND

C CROTCH
COMFORT

NE SUPPORT

Healthknit
REG. U.S. PAT. OFF.
FALL-WINTER

MacDEE
ANKLE LENGTH
SIZE 42
EXTRA LONG

Cantilever Action
GIVES GENTLE LIFT

"No fit like Healthknit"

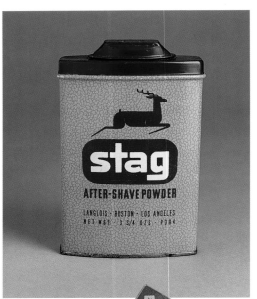

Rexall Drug Stores sold Stag after-shave powder in the mid forties. The nicely stylized buck logo reflects the angles and curves of the product's logotype underneath.

This counter display from the mid fifties with its handsome graphics and clever structural design attempts to advertise a "new" Norelco Speedshaver that has distinctly older, thirties streamline styling.

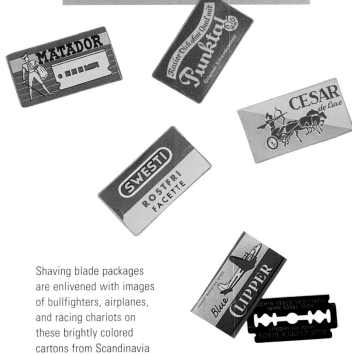

Shaving blade packages are enlivened with images of bullfighters, airplanes, and racing chariots on these brightly colored cartons from Scandinavia and Germany, c. 1950.

82

Unlike today's condom packaging with blantant images of couples caressing, these small tins and paperboard containers from the forties and fifties, their contents made of genuine latex, used the discreet approach that ran from the storytelling (Texide), to the elegantly minimal (Trojans and Sheik), to the humorous (Spares).

record
keeping

In the 1940s, enjoying your favorite Duke Ellington jazz composition or Xavier Cugat rhumba on the phonograph was a lot of work. Before the introduction of the long-playing vinyl record—the LP—in 1950, an album was just that, a collection of three to five records, each playing only four to five minutes per side.

These early records were heavy, shellac-coated, and fragile. Sturdy packaging was therefore de rigueur. A set of records was stored in kraft-paper sleeves that were bound and sandwiched between pasteboard covers. Wrapped in dark-colored paper and stamped in gold with the name of the record, they earned the moniker "tombstones."

Some record companies had experimented with design in the mid thirties to boost Depression-era sales, but it wasn't until the forties that things really got spinning. With supermarkets popping up all over the country in the mid forties, Americans no longer wanted to shop for goods displayed behind counters. The era of impulse buying had begun. In 1945, self-selection was introduced in record stores. Placed on a rack with the album cover facing the potential buyer, the understated, tombstone packaging was now obviously out of tune in this new marketing environment of colorful, eye-catching graphics.

A new breed of package designer was born overnight—the album-cover designer. Alex Steinweiss is often credited as the father of record-jacket design. Creating a visual identity for Columbia Records, he was a prolific artist who painted in the flat, colorful style of European poster designers such as Joseph Binder. Other high-note album designers of the time were Jim Flora, Bob Jones, and David Stone Martin.

By the mid fifties, a new generation of art directors working with jazz and rock 'n' roll musicians preferred the modern look of photography over illustration. Different styles of music and consumers' rapidly changing tastes in the fifties and sixties closed a chapter on what now is considered the classic age of the illustrated album cover.

Latin music and dances were all the rage in the United States in the forties and fifties, as was the Spaniard Xavier Cugat, the world-acclaimed "Rhumba King." Holding four, shellac-coated 78 rpm records, *Rhumba with Cugat* was illustrated by Columbia Recording Studio's top designer, Alex Steinweiss, c. 1947.

Andre Kostelanetz conducts the Philhamonic-Symphony Orchestra of New York playing Gershwin in Concerto in F and his own orchestra preforming Ferde Grofé's Grand Canyon Suite. Although the album cover for the Grand Canyon Suite is attractively executed, the more imaginative cover for Concerto in F has typography that reads like scintilating musical notes or a constructivist poem. Both albums were released by Columbia Records in the mid forties.

Ferde Grofé's
GRAND CANYON SUITE
ANDRE KOSTELANETZ
and his orchestra set M-463
COLUMBIA MASTERWORKS

duke ellington and his orchestra

on a turquoise cloud
hy'a sue
new york city blues
lady of the lavender mist
golden cress
the clothed woman
three cent stamp
progressive gavotte

mood ellington

❸ set c-164

columbia records

The influence of renowned Modernist poster designer Joseph Binder can be seen in these Frankie Carle albums by Alex Steinweiss. Steinweiss assisted Binder for almost three years in the late thirties where he learned to master airbrush illustration.

This dramatic cover for *Mood Ellington*, created by Bob Jones, was recorded by Duke Ellington and his orchestra in 1948. Black bars, evoking the image of piano keys, dance on a background of purple haze.

Bob Jones designed Woody Herman and his Woodchoppers' cover and Alex Steinweiss illustrated *Continental Tango* conducted by Marek Weber while both men worked for Columbia Records in the late forties. Colorful stripes were used as eye-catching backgrounds for all types of packaging throughout the fifties.

A West Coast Jazz Anthology, dated 1956, featured such jazz greats as Gerry Mulligan and Chet Baker. This early LP (long-playing record) has a cover that owes much to the cool geometrical precision of the Modern painter Piet Mondrian. Robert Parent's silhouetted photograph of the group playing is blurred almost to abstraction.

MODERN JAZZ

a west coast jazz
anthology

WITH GERRY MULLIGAN, CHET BAKER,
BUD SHANK, CHICO HAMILTON AND OTHERS

Picture frames were widely used in the forties and fifties as a playful, surreal design element, or to literally "frame" an image. Wood-textured borders contain couples dancing on the cover of these *Arthur Murray Favorites* albums by Capitol Records. The rhumba and mambo were popular dances as were the fox trot and waltz during this time. Both LPs played at a speed of 33 1/3 rpm and date from the mid fifties.

Stan Kenton's two albums for Capitol Records feature stabbing, criss-crossing linework on their covers. *Encores* is unabashedly surrealist inspired, while *New Concepts* omits a slice of red and thereby highlights a black and white photograph of Kenton's eyes. RCA let Jim Flora's distinctive illustration style sell the cover for *Collaboration*, Shorty Rogers and Andre Previn's 1955 album.

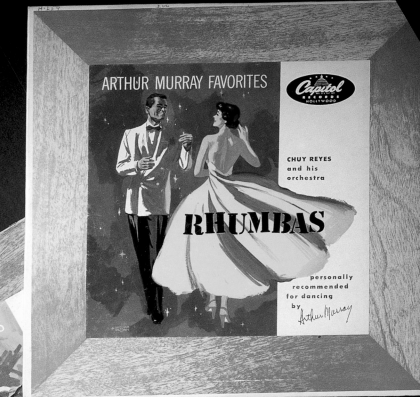

ARTHUR MURRAY FAVORITES

Capitol RECORDS HOLLYWOOD

CHUY REYES and his orchestra

RHUMBAS

personally recommended for dancing by

Arthur Murray

ARTHUR MURRAY FAVORITES

MAMBOS

personally recommended for dancing by

Arthur Murray

The Remco Electronic Space Gun, c. 1950, sold in a carton whose lid could be folded back for store display. The battery-operated toy didn't really do much, but a diagram on the carton's bottom panel outlined some pretty exciting features: "color ray selector turret, knurled non slip hand grip, atom chamber, nuclear exhaust ports and tele-video sight."

kid
stuff

By the end of the 1940s, three million television sets glowed from America's living rooms, although there wasn't much to watch in the early days of television. That all changed with the premiere of the *Howdy Doody Show* in 1947. Television's first program for children featured the fringed Buffalo Bob and his puppet sidekick, the show's namesake.

The *Howdy Doody Show* was a big hit with kids, and it didn't take long for television producers and sponsors to exploit this untapped merchandising gold mine. Networks that broadcast *Howdy Doody* and other popular children's shows from the forties, fifties, and sixties—the golden age of television—collected handsome licensing fees from manufacturers who marketed products using the programs' characters. Capitalizing on a show's popularity, product spin-offs seemed limitless—toys, board games, clothes, trading cards, food products, lunch boxes, and more.

When color televisions replaced the older black and white sets, kids expected the packaging of their toys and games to be as vivid and exciting on the store shelves as it was on the screen. Package designers used every trick in the book to sell to these hard-to-please, pint-sized consumers. Illustration on the boxes usually won out over photography. Exaggerating the product's size or playing action was accomplished with airbrushing. Colors could be brightened and sweeping lines added to a model Chevy's back fins, suggesting that a fast rod was parked inside. Consumer testing dictated shades of pink and violet packaging for girl's products, while boys seemed to be drawn to red and orange.

Vintage toys and games from the *Howdy Doody* and *Captain Kangaroo* era are avidly purchased today. Prices on these collectibles have skyrocketed as adult baby boomers restock their childhood toy boxes. Packaging is important to collectors, too; the toy or game is worth top dollar if still MIB: Mint In Box.

Contracting to build, contructing, and then orbiting rockets was the main objective of playing Astronaut (left). The overwrap on Astronaut the New Game of Outer Space is lithographed with a realistically rendered space suit of 1962 vintage worn by a determined-looking spaceman.

Space-O (far right) was another card game that capitalized on America's new infatuation with satellites and space travel. Arrco Playing Card Co. of Chicago, Illinois, manufactured this game in the late fifties "for boys and girls from 4 to 10. . .and just about everybody."

ASTRONAUT

THE NEW GAME OF OUTER SPACE

EXCITING! TAKES YOU FROM BASE TO SPACE!

A GYROSCOPIC AIRFOIL · PRESEN

FLYIN-

It flies ★ STRAIGH
★ CURVES
★ CIRCLES
★ SKIPS

WILL NOT BREAK

Invented by Fred Morrison in the late fifties, Flyin'-Saucer (right) cashed in on a time of great interest in UFOs and outer space, and it also was a lot of fun to play. Wham-O bought the rights to his patent and in 1958 renamed his plastic "pie pan" Frisbee.

This beautifully lithographed, pressed-metal flying saucer toy (far right) came packaged in a cheaply fabricated, set-up box with a simple glued-on label. "No. 562" and "Made in Western-Germany" are the only clues to its identity, c. 1950.

At a time when "Made in Japan" meant "cheaply made," cowboys, Indians, and western imagery were the hottest things going (with the possible exception of astronauts, Martian invaders and outer space) for school-age kids. The card holding the cowboy watch and sheriff's star badge did double-duty as both packaging and as another toy—a black mask. Western Set also held a toy watch and a badge of criss-crossing guns. Both packaging and toys made in Japan for the American market, c. 1950.

A cowboy's lasso frames a center die-cut allowing the buyer to see part of the assortment of color pencils. Eberhard Faber Pencil Co. of Brooklyn, New York, retailed their academic product with exiting images of the Old West, c. 1950.

A crudely drawn, hard-riding cowboy advertises Texas Trail Treachery in his lasso on the cover of this fifties, 8mm home-movie carton.

Measuring under two inches in length, Wild West's half-pint box and cards are none-the-less richly illus-trated with Indian scouts, bronco busters, bull doggers, and singing cowboys.

The fifties and sixties are considered by collectors to be the golden age of plastic kit models, with model cars usually winning at the cash register. Lew Glaser, founder of Revell Toys, marketed the first easy-to-assemble kits in 1951. Designed by Jake Millard Williams Associates, Pyro Plastics Corporation launched Design-A-Car around 1960. Theoretically, one could "design and build over 25,000 different models!" using Pyro's "computer"—a paperboard, sliderule-type calculator. The bright-colored vehicle shown on the cover borrows the 1959 Chevy Impala's "gull-wing" tail fins.

While little boys were getting kits to design automobiles and space rockets, little girls were getting slightly less challenging toys such as Let's Do the Dishes. Since women (and minorities) had greatly diminished career expectations in the fifties, teaching them "how fun it is to wash the dishes" at a tender age seems a step in the right direction for proper social conditioning. Manufactured by Amsco of Hatboro, Pennsylvania, in the mid fifties.

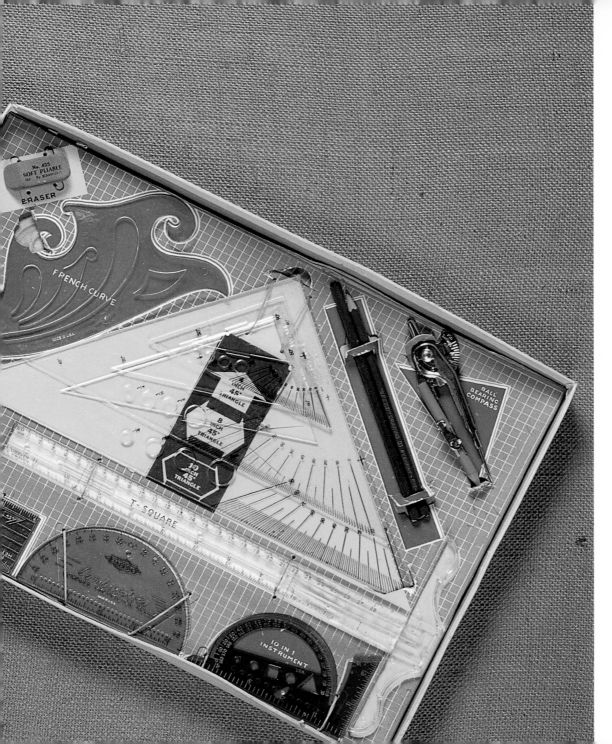

Science and the future were hot marketing items to children in the fifties and early sixties. Containing "precision instruments for young engineers of the future," the allure of designing space rockets for the city of tomorrow is the sales message on the fully illustrated cover of Space Scientist Drafting Set, manufactured around 1958 by Hassenfeid Bros., Inc., of Central Falls, Rhode Island.

Quiz and game shows have been popular television fodder since the early days of broadcasting. The cover for You Don't Say!, based on the NBC television game, shows two happy couples gaming with play money. As with most Milton Bradley games, this one is "for ages 10 to adult," dated 1963.

"As seen on NBC television," this home game of Concentration closely mimicked the screen version, dated 1958. Players rival to win gifts and money by uncovering matching pairs of cards, which are then removed from the board to reveal parts of a hidden rebus or word. Sixty different puzzles are contained on a scroll inside the plastic Rolomatic Puzzle Changer. What's the answer to this puzzle?—"cash and carry"

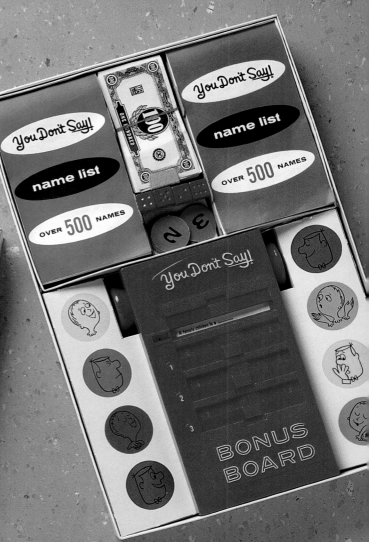

Park and Shop has got to be the quintessential living-in-suburbia board game. Hustle and bustle rule in this traffic game by Milton Bradley, dated 1960. After each player picks a parking spot, the object of the game is to shop the fastest and the first one home wins.

How to Succeed in Business would probably be called How to Be Sued for Blatant Sexism if sold today. Another Milton Bradley game dated 1963, How to Succeed's box cover says it all with the illustration of a leggy secretary and a phone-holding CEO extending out of the sides of the same black office chair.

Eddie Cantor's GAME

'TELL IT TO THE JUDGE'

REGISTERED IN U.S. PATENT OFFICE

BACK SEAT DRIVER

7 2

Parker Brothers Inc.
SALEM, MASSACHUSETTS · NEW YORK · SAN FRANCISCO · CHICAGO
MADE IN U.S.A.

Tycoon's box cover features a funny, free-wheeling illustration of a fat-cat, cigar-smoking, stock broker whose specs form the double *o*'s in the board game's name. The artwork uses jagged, abstract forms of white and gray for no other reason but to add excitement and draw attention. The wonderfully off-kilter box design was created by Games, Inc. of Minneapolis, Minnesota, in the mid fifties.

for children over 10 and for fun-loving adults

CONTAINS ALL THE ACTION AND SUSPENSE OF STOCK MARKET SPECULATION, MARKET REPORTS, HIGH FINANCE

Parker Brothers Inc. released the board game "Tell It to the Judge" in the mid fifties, coat-tailing the popularity of comedian Eddie Cantor. The boxed game's cover is a superb example of a fifties illustration style: the combination of black, rough-lined drawing, "off-register" color, and the stylization of objects rendered in flat color.

canned laughter

A printed label wrapped around a fifteen-ounce can hatched a souvenir of Los Angeles, California. It contained "hydrocarbons, nitrogen oxides, sulfur dioxide, organic oxides, aldehydes, and formaldehydes." Recipients of this gag gift were advised "to insure freshness and purity keep container tightly sealed." Packed for Los Angeles Smog Corporation, c. 1960.

Given as a present at a birthday or bachelor party, the gag gift provoked unpleasant and embarrassed reactions from the recipient. Compared with the decades before and after, the fifties had to be the high point in the history of this type of low-brow humor. Whether we consider any of these cultural artifacts truly funny today is irrelevant. The aggressive, in-your-face humor of our contemporary, late-night talk hosts probably would have been considered crude and offensive in the age of Jack Benny and Bing Crosby.

The fifties gag gift was a cheap, middle-class guffaw usually packaged in an inexpensively printed set-up box. The set-up box is sturdier than the folding carton because it is rigidly constructed with reinforced corners. When over-wrapped in costly papers or cloth, set-up boxes are commonly used for gift or high-end items such as perfumes or specialty products. Die-cut inserts inside the box often form platforms to hold the contents in place (page 77). For the cheaply manufactured gag gift, the container's lid was wrapped with a low-grade paper printed in one or two colors.

Like any practical joke, the fifties gag gift relied on the impact of the one-liner. Subtlety was not the goal. The humor was usually vulgar and played to scatological and sexual themes. In a period of political and social conservativism and repressive sexual mores, this form of humor offered a release valve for tension, insecurities, shyness, aggression, and frustration. It often exploited ethnic and racial stereotypes, revealing America's unease and suspicion of immigrants. For the lower classes, the nervous laugh that the gag gift elicited, helped defuse the frustrations of living with the demands of fifties conformity.

The gag gift was typically bestowed at stag parties, house warmings, cocktail parties, and as a souvenir from a recent trip. Visitors to Southern California couldn't bring back the sunshine, but they could bring back a can of "genuine" Los Angeles smog that surely would have been passed around to titters and smirks even at the most sedate Tupperware party.

Here's an eye opener!

Puns are the copy of choice on Korrect-Way Displays' folding card that holds a girlie "eye opener" give-away. Your "mail" instincts will be aroused when you use her to open your letters. The amber-colored, plastic letter opener is disrobed as you pull her out of her paperboard party dress. The ball balancing on her head puns: "Seen in the best circles." This advertising reminder for Beau Monde Manikins was manufactured in St. Louis, Missouri, in the late forties.

An inexpensive, chip board set-up box holds a pair of B.S. Protectors—transparent acetate cups strung together with elastic strings. As the illustration inside demonstrates, fitting these cups over your ears blocked out bragging friends, nagging wives, loud-mouth bosses, and neighbors who holler profusely. A timid toreador and charging bull illustrate the lid, c. 1955.

Meet Ellen, the Eye Opener, a girl who'll appeal to your "mail" instincts . . . as she opens your mail, let her serve you to remind you of Beau Monde Manikins . . . the figures you should really get acquainted with . . . the ones that caused the greatest "upset" in manikin history . . . and will continue to be the "manikins of fashion significance".

KORRECT-WAY DISPLAYS
Div. of American Fixture & Mfg. Co.
2300 LOCUST ST.
ST. LOUIS 3, MO.

beau monde
MANIKINS WITH MAGNETISM
Everywhere—

What would a bachelor's party be without off-color humor? The punchline of this gag gift (left) needs no further elaboration. The lithographed coverwrap adds a light-hearted touch to the set-up box, c. 1950.

The perfect gag gift for the cheapskate friend (below) who claims that he's only being frugal when he doesn't tip the waiter or bellboy. Let him "sock away" his pennies in this purse. The image of the stereotypical "thrifty Scotsman" was conscripted by companies in the forties and fifties to sell a wide assortment of products, c. 1950.

bibliography

"America's Amazing New Easy Foods." *Look Magazine* Vol. 1, no.15 (January 6, 1959): 64–65.

Battersby, Martin. *The Decorative Thirties.* New York: Billboard Publications, 1988.

Bender, May. *Package Design and Social Change.* New York: AMACOM, 1975.

Clark, Hyla M. *The Tin Can Book.* New York: New American Library/Times Mirror, 1977.

Erbe, Maureen. *Made in Japan: Transitor Radios of the 1950s and 1960s.* San Francisco: Chronicle Books, 1993.

"The Fabulous Market for Food." *Fortune Magazine* Vol. XLVIII, no. 4 (October 1953): 135–39.

Fleming, William. *Arts and Ideas.* New York: Holt, Rinehart and Winston, 1968.

Franken, Richard B. and Carroll B. Larrabee, *Packages That Sell.* New York: Harper and Brothers Publishers, 1928.

Frey, Tom. *Toy Bop: Kid Classics of the '50s & '60s.* Murrysville, Penn.: Fuzzy Dice Publications, 1994.

Gray, Milner. *Package Design.* London: Studio Publications, 1955.

Herdeg, Walter. *Packaging.* Zurich: Graphis Press, 1959.

——*Packaging 3.* Zurich: Graphis Press, 1977.

Hine, Thomas. *Populuxe.* New York: Alfred A. Knopf, 1987.

——*The Total Package.* New York: Little, Brown and Company, 1995.

"How Kids Earn Their Keep: Marketing to Children." *Print Magazine* XLVI:VI (November/December 1992): 54–57.

"Industry/The Packaging War." *Time Magazine* Vol. LXXXI, no. 18 (May 3, 1963): 96.

Kamekura, Yusaku. *Trademark Designs of the World.* New York: Dover Publications, 1981.

Milton, Howard. *Packaging Design.* London: Design Council, 1991.

Mingo, Jack and John Javna, *The Whole Pop Catalog.* New York: Avon Books, 1991.

Morgan, Hal. *Symbols of America.* New York: Viking Penguin, 1986.

Neubauer, Robert G. *Packaging the Contemporary Media.* New York: Van Nostrand Reinhold, 1973.

Opie, Robert. *The Art of the Label.* Secaucus, N. J.: Chartwell Books, 1987

Pulos, Arthur J. *The American Design Adventure.* Cambridge, Mass.: MIT Press, 1990.

Remington, R. Roger. *Nine Pioneers in American Graphic Design.* Cambridge, Mass.: MIT Press, 1989.

Sacharow, Stanley. *The Package as a Marketing Tool.* Radnor, Penn.: Chilton Book Company, 1982.

Sutnar, Ladislav. *Package Design: The Force of Visual Selling.* New York: Arts, 1953.

ailfins   alvin lustig   sputnik   slinky

tv dinners   korean war   thelonius monk

eisenhower   saul bass   abstract expressionism

boomerang   color tv   ed sullivan   jazz

plastics   jean arp   beatnik   buddy holly

civil rights   howdy doody   andy warhol

lester beall   jack kerouac   surrealism   kool-aid

supermarkets   barbie   atomic bomb   elvis

cool   pop art   game shows   the beatles

westerns   walter paepcke   paul rand

tapering   william golden   biomorphic